Living Off the Land
A Beginner's Guide to Being Self-sufficient

By
Darla Noble

Mendon Cottage Books

JD-Biz Publishing

Table of Contents

Introduction

There's a little bit of pioneer spirit in all of us. We can't help it...it's in our blood. For some this pioneer spirit shows itself in someone's determination to climb to the top of the corporate ladder. But for others, this pioneer spirit takes them back to their roots...literally; giving them the desire to be self-sufficient to the greatest extent possible.

If you are reading this book you are most likely someone wanting to be more self-sufficient. Good for you! With the cost of food going higher and higher every week (literally) and the nearly-constant revelations of the negative effects of chemicals, processed foods and other things we ingest, it's a shame more people aren't willing to do more to get back to the basics of providing for themselves.

Yes, it's true you may raise a few eyebrows or be asked if you're hooked on reruns of "Little House on the Prairie", but that's okay. Besides, I bet those same people will be wishing they would have been a little less critical when you're giving away excess produce.

Anyway...the purpose of this book is to give you the direction and encouragement you need to be able to be as self-sufficient as possible. So without further ado...let's get started.

Chapter 1: It's What You Do With What You've Got That Matters

Before we really get started on what it takes to be able to be self-sufficient, let's take a minute or two to go over a few words or phrases you'll see repeatedly and define what is meant by each of them.

1—Self-sufficient: To be responsible for your existence to the greatest degree possible. NOTE: This is a relative term, however, as geography, zoning and other issues play a role in how self-sufficient we can be.

2—Farm: It doesn't matter whether you have two acres or two hundred acres…if you are harvesting what your land produces for the purpose of sustaining yourself, your family and/or others, you have a farm.

3—Schedule F: The tax form you must file with the IRS in order to take the tax deductions due you IF you have farm INCOME. NOTE: Not profit, but income.

4—Stocking rate: This is the number of animals your land can support or sustain without harming the land under normal conditions. To figure the stocking rate of your land, you need to do the following: 1) Figure how much forage (grass/hay) the land you intend to graze will produce (in pounds). To do this, determine how many bales of hay each acre of ground would produce and the poundage of those bales. 2) Determine the average weight of the species of livestock you wish to graze. Divide this number by .25. This gives you the amount in pounds each animal needs per year. You can then figure how many animals you can raise on an acre of ground.

NOTE: If you intend to graze more than one species of animal (which you probably do), use the largest animal for this calculation.

Example: If one acre of land of mixed grasses yields on average of five 4x6 bales of hay weighing 800 to 900 pounds OR seventy to eighty 70 pound square bales, you will have between 4,000 and 4,500 pounds of hay or grass for your livestock to eat per acre. This means that if you have two acres of ground to graze or mow for hay, you will have between 8,000 and 9,000 pounds of forage available. Using this information you can now figure how many animals your farm will support.

5—Rotational grazing: Moving your livestock from pasture to pasture for the purpose of grazing to promote parasite control and proper grass management practices.

Image of Schedule F (IRS)

Now that that's done, let's get down to the business of preparing you to be as self-sufficient as possible. If I had to sum it up in one word, I think that word would be management. How well you manage your farm (no matter how big or small) determines how successful it is (or isn't).

Management of your farm determines just about everything. It determines how much you can successfully raise and harvest off your land. It determines how healthy your livestock is. It determines whether your produce is ridden with insects or healthy and delicious to eat. And finally, management determines whether or not your initial energy and efforts pay off or are a waste of time, energy and money.

Proper management includes keeping your pastures in good shape.

But what is good management when it comes to having a farm that allows you to be self-sufficient? Good management means you are:

Sufficiently prepared. The number of people who have jumped into small farming for the purpose of being self-sufficient in a state of unpreparedness is a big one. I know this to be true because I have personally witnessed a large number of these train wrecks (that's putting it mildly). This is a sad state of affairs; leaving people disillusioned about the agricultural industry, with a lowered self-esteem and often with a bank account depleted of much-needed funds.

Once you decide to adopt a lifestyle of self-sufficiency, it is important that you follow the advice you probably heard more than once from your grandmother: "Anything worth doing is worth doing right". In this case, "doing it right" means to be plan and prepare yourself and your land for what you plan to do. This means having adequate and proper fencing in place, having easy access to water in different locations on your land, owning the proper equipment to do what you intend to do and having adequate housing, storage and/or housing for your livestock and products.

Hard work pays off

Willing to work. I'd be willing to bet you've also your grandma say something to the effect that a little hard work never hurt anyone. She was right on that one, too. Becoming even slightly self-sufficient takes effort and energy on your part. Gardens don't plant, tend and harvest themselves you know. Likewise, cows, chickens and sheep don't completely feed and tend to themselves. Being a good manager means you must be willing to work for what you have. And let's just say upfront that the work isn't easy. No, it's not easy, but it is rewarding and great physical activity and exercise.

Adequately knowledgeable. This by no means implies that you need to be an expert gardener or know everything there is to know about raising whatever types of livestock you choose to raise. What it *does* mean, however, is that you need to have a working knowledge of what you are growing/raising as well as being willing and able to learn more along the way. Learning along the way may happen just by doing (experience is a great teacher, you know), as well as networking with other producers and attending events that enable you to learn.

Assuming all of this is so, you will need to take your management skills to the next level. This means staying on top of weeding, feeding, harvesting and processing your flowers and veggies and giving proper care to your livestock in order to keep them in good health and at the top of their game in regards to productivity.

Proper management means you understand and take advantage of concepts like vertical growing, rotational grazing, co-grazing/growing, solar energy, frugality and thriftiness and optimal utilization of your property.

Chapter 2: Moo, Bah, Cluck And Oink

No problem figuring out what this chapter is about, is there? That's because the livestock on your farm play a major role in the 'game' of being self-sufficient and we need to make sure you have as much information as we can cram into this chapter to assist you in your quest to be as self-sufficient as possible. I'm not going to spend a lot of time being technical or going into too much detail regarding the specific care of a particular species. Instead, I'm going to deal more with the general rules of successfully raising livestock in order to be self-sufficient...even on small acreages.

Rule #1. Less is definitely more realistic and sensible when it comes to the number of animals on your farm. This doesn't necessarily mean the number of different varieties, but rather the number of any one type of animal. Confused? Let me explain....

Let's say you have a five acre plot of land. If the stocking rate (see chapter 1) says you can support a milk cow and two steers or twenty sheep, you'd be better to be cautious and hold yourself to a milk cow and a steer OR a steer and a few sheep and a milk goat or two in order to keep your ground productive and sustainable.

Rule #2. Being self-sustaining doesn't put you in the livestock production business. Don't waste your space on keeping a bull or ram for breeding once a year. Instead, borrow a bull or ram or pay for siring services if you are going to have a milk cow or milk goats. Otherwise, just buy to weaned calves, lambs or pigs to raise and feed out for butchering.

Raising a steer for next year's meat

A few lambs for butchering

NOTE: The exception to this is chickens. A chicken house with a couple of dozen hens and a rooster is completely feasible and manageable both for having fresh eggs and for raising chickens for butchering if you so choose.

Laying Hens

Rule #3. Treat your land right. This means fertilizing at least every other year, not allowing animals to eat the grass to the ground; killing it out and not over-grazing it with too many animals. It also means keeping noxious grasses and weeds out of your pastures and garden areas.

Rule #4. Low-maintenance doesn't mean no maintenance. Some people, in the name of self-sufficiency or naturalistic living, plop a few animals on their land and do nothing else until it's time to butcher them and when they do, they are dissatisfied with both the quality and quantity of the meat they end up with. My question to them is this: What did you expect? If you don't feed your children they won't grow. If you don't water your garden it won't produce. If you don't feed your livestock they won't 'give' you a quality carcass for your freezer.

This no maintenance frame of mind also threatens the health of livestock belonging to people who want to 'go green' or be 'all natural'. While I'm not saying there is anything wrong with using as few medications and chemicals as possible, there sometimes comes a point when these things are needed. NOTE: This can be especially true when you are farming on a small scale and animals are confined more than they would be in large operations.

A healthy flock of sheep in a low-maintenance setting

Rule #5. Be connected. By being connected, I mean you need to make use of the resources available to you for learning how to properly care for your animals and how to get the most out of your land without depleting it into nothingness. In working with small numbers of livestock, you'll simply not have the opportunity to become an expert (if there is such a thing) on sheep, cattle, goats or whatever it is you are raising. That's okay! You don't have to be an expert—you just have to be willing to network and connect with other small farmers to share information, experiences and knowledge.

Even the youngest members of the family can network. Rule #6. Have a plan. I cannot tell you the number of people I've talked to over the years who've had big plans for adopting a self-sufficient lifestyle—a 'big picture' sort of plan with no thought to details. Let me give you just one such example…

A young couple who attended an annual farm show we did stopped to talk to us about purchasing sheep for their small farm. They were asking the right questions and seemed to have a grasp on what they were doing. But as the conversation progressed, we learned that this couple had eight acres on which they planned to house two dozen hens, a pig they would purchase each year to feed out and butcher, a steer they would purchase each year to feed out and butcher, six milk goats and a billy goat, a dozen ewes and a ram, a few bee hives, an acre of sweet corn and another half-acre garden.

Their intent was to raise and grow enough for themselves as well as some to sell to supplement their income. This, too, is a good idea…if you've put together a plan that makes sense. This couple didn't understand that they couldn't throw all the animals in together and just let nature take its course. They didn't understand that milk goats would need a period of rest (non-lactation) before giving birth each year. They didn't understand what it took to plant, cultivate and harvest an entire acre of sweet corn—especially since they didn't have some of the equipment they needed to avoid doing it all by hand.

This couple's general plan or idea to become more self-sufficient *and* supplement their income by raising and producing a variety of crops was a good one. They just hadn't taken the time to be educated on what would and wouldn't work. But because they were sincere in their desire to make the most of their small farm, they were willing to listen to some sound advice to tweak their plan to make it a workable one.

The lesson to be learned here is not only to have a plan, but consult with the right people to make sure your plan will work.

Rule #7. Make sure your land has no restrictions. As crazy as it may sound, there are rural communities that have some restrictions on what can and cannot be done with your land. You will probably find that most restrictions will be on having pigs or hogs or you may have to prove you have adequately planned for erosion prevention.

Rule #8. Layout is everything. How you lay out or arrange your barn and pastures is essential when it comes to making the most efficient use of your land.

When thinking about the layout of your farm, you need to take the following into consideration:

The barn. You will need a barn that can house your livestock when/if necessary, one in which you can store hay for winter feeding and one in which you can access from every pasture. One that is a 30x50 ft. is adequate for being able to use for a milk cow, steer and a dozen or so sheep.

Your barn should have a walk-through door being accessible from outside the fence. On this same side a smaller feed and tool shed should be placed nearby. The barn should have two 15 foot roof to ground sliding doors (one on each end) that can remain open for access and ventilation. These doors should have gates that open IN to hold livestock if necessary. You should also partition the barn with cattle panels to suit your needs (a milk stall, lambing pens, working/loading alley, etc.).

The barn needs to accommodate the different species of livestock separately if necessary.

Outbuildings. While it needn't be large, having an additional storage shed (even one of the pre-fab ones you see at the farm store) is a good idea. You can use it to house your mower, tiller, gardening tools, feed and livestock supplies such as feeders, halters, vet supplies, etc.. You can also add a covered lean-to to either this building or the barn to house your tractor. This building should be outside the perimeter fence so to allow for quick and easy access from your yard/garden.

Chicken house. If chickens are to be part of your self-sufficient lifestyle, you will need a small chicken house. The chicken house can be located near the barn and storage building OR it can even be included as part of the barn, but the chickens should not be with the other livestock.

Chickens will need a fenced area and a ramp or 'run' to come and go from their shelter at will. A chicken house will need to have electricity for lights (to keep them laying when the days get shorter) and to keep their water from freezing. Your chicken house will need to be one that is easy to clean since manure and ammonia build-up happens quickly and can affect the productivity of your hens.

Pastures. Even if you have as little as a couple of acres, it is advisable to section off your pasture into 1-3 acre sections to allow for rotational grazing.

Late winter pasture

NOTE: Sheep or goats and cows can easily co-graze, so it is entirely practical to include a gate in the fence that divides your two main parcels of pasture. You will, however, find it best to keep the different species separate during the time they are having their young.

Additionally, since most self-sufficient farms include a milk cow(s) or goat(s), you will find it much easier to keep these animals separate in order to be able to call them up to milk and feed separately from everything else.

A place to bring your milk cow into eat/milk

While it should go without saying that planning your pasture should include planning your fencing, well…I've learned over the years to never assume anything when it comes to talking to people about farming. So let's spend a few minutes on how to properly fence your farm.

Perimeter fencing should be able to a) hold any form of livestock in and b) keep predators out. In order to fulfill both of these requirements, I strongly advise using woven wire fencing for your perimeter fencing. You should also have one or two gates large enough to drive through with a truck and livestock trailer for loading and unloading livestock and at least one of the gates needs to be accessible no matter what the weather or ground conditions are.

Once you have your perimeter fencing in place, you can then put in your cross-fencing. Cross-fencing can either be permanent or as temporary as electrified fencing on spring clips you can move at will to increase or decrease the size of your pastures.

Water. A readily-available water source is a must on your farm. Installing a frost-proof waterline and hydrant or two just outside the barn (and outside the pastures), just close enough to the fence to reach water tanks easily with a short hose, is also one of those things you can't do without.

Making this extension from your well or another outside water source is not difficult and relatively inexpensive…especially when you consider the time, energy and physical labor it will save you. Trust me on this one. We spent one very looooooong winter hauling water for some calves from the house to a barn on an adjoining property to ours. We knew it wasn't an ideal situation but because of other circumstances beyond our control it became necessary. It was not fun.

Are you beginning to get the message—that it really does come down to good and proper management? It doesn't take a degree in animal science or agriculture. It doesn't take generations of farmers in your family. It doesn't take lots and lots of land. It takes…say it with me…management.

Chapter 3: Eat Your Veggies

In this chapter we are going to talk about getting down and dirty…literally. Some people wanting to live a self-sufficient lifestyle may or may not be able to include livestock in their plan of action for any number of reasons. But growing your how fruits and veggies…well that's a different story altogether. Growing your own fruits and vegetables is something virtually anyone can do to some extent or another.

Planning a garden is a bit like sitting down to Thanksgiving dinner. There's so much to choose from it's difficult to resist putting more in your garden than you can grow, take care of and eat. So in getting ready to plant, remember…

Gardening is rewarding, but time-consuming work. Breaking ground for a garden and preparing the soil for planting is just the beginning. Once the seeds are planted, you must water the plants, feed the soil, keep it free of weeds and pests and be ready to harvest what you grow. Once harvested, you also need to have the time to preserve it (can or freeze).

A bountiful garden provides plenty for a large family.

I don't say this to deter you, I say this to keep you from being overwhelmed and ending up feeling like you've failed. So rather than taking on more than you have the time, desire and resources to manage, take a few minutes to sit down and really think about what you should plant. Ask yourself:

HOW MUCH ROOM DO I REALLY NEED? There is no set answer to this question. The amount of room you need will be determined by what you grow and how much you grow. Corn, melons, cucumbers and squash require quite a bit of room, while beans, peas, tomatoes, potatoes, and greens require considerably less room.

HOW MUCH DO I REALLY NEED? If you are a family of six and you all like green beans, then make sure you plant enough green beans to eat fresh and can or freeze for serving twice a week to your family. Are you the only one who likes pickles? Don't waste your time, space, money and energy canning more than a half dozen or so pints of pickles. Squash doesn't freeze or can well, so don't plant any more than your family will eat during the summer season. If you plan on storing potatoes to use throughout the winter, make sure you have a cool, dark, dry place to store them. If not, you'd be better off canning them. In other words, plant the foods your family will eat in the quantities necessary to feed them for a year at rates you would normally feed them.

Fresh tomatoes

If you *do* have the room, time and energy to grow more than you need, you can supplement your income by selling produce at the local farmer's market. No business license is necessary to do so, the booth fees are usually very inexpensive and it's a great way to meet other growers.

WHAT SHOULD I GROW? There's nothing self-sufficient in growing things you won't eat. You should grow what you and your family like to eat and (at the risk of repeating myself) grow enough of it to take you through the winter.

The following is a list of vegetables that can be canned and/or frozen to preserve for winter use: Corn, Green beans, Potatoes, Tomatoes, Peppers, Okra, Onions, Peas, Cabbage, Pickles, Beets, Carrots, Broccoli, Cauliflower and Sweet potatoes. You can also cook pumpkin down and freeze or can it. And finally…asparagus can be frozen but unless flash-frozen, it is a bit mushy when cooked. Asparagus is also a vegetable that takes a while to become established and must be grown in a place that can remain undisturbed. If asparagus is for you, you will need to designate one end of your garden as an asparagus bed and not touch it with the plow or tiller.

NOTE: Some people separate the asparagus bed from the rest of the garden with a compost pile. Dumping peelings and other food scraps, leaves and grass clippings in a compost pile on the edge of your garden makes it easy to add the compost to your garden for the purpose of enriching the soil.

Another important factor to keep in mind: When planning how many tomato plants to plant, take into consideration all the things you plan to use them for. For example, you will need enough tomatoes for making sauce, salsa, chili sauce, etc.—whatever your family likes. The same holds true for sweet potatoes. You can store the like you do white potatoes and you can cook/mash them and freeze them for making sweet potato pie, etc.

Here is a list of vegetables you cannot can or freeze: Squash, Zucchini, Greens, Eggplant, Melon and Lettuce. You will only need to grow enough of these to consume in-season and a little extra to share with friends and neighbors or sell at the farmer's market, if you so choose.

NOTE: Zucchini can be grated and frozen for making muffins and breads, but must be squeezed and drained after thawing before using.

IS MY GARDEN A HEALTHY GARDEN? The best garden is a healthy garden. In order to make it healthy, the soil needs to be filled with nutrients, have good aeration, kept free of weeds, and receive adequate water and sunshine.

Where you put your garden will also play a role in how well it produces for you. You need to make sure your garden will drain well; meaning the water runs away from the garden rather than pooling in the garden creating a swamp land. Gardens that receive the afternoon sun rather than the morning sun tend to do better as well.

To add nutrients in the soil, working peat moss and/or into the soil never hurts. If the compost is kept at the edge of the garden as was mentioned earlier, it will be handy for you to work into the soil. Dried manure from rabbits, chickens, cows, sheep or llamas also works very

well. Of all those listed, however, rabbit manure is by far the BEST for you garden.
I realize, however, that in being self-sufficient you will be more likely to have cow, sheep or chicken manure; manure that cost nothing more than the manual labor it takes to get it from the barn lot or chicken house to the garden. By all means, use it! Just be aware of the fact that cow and sheep manure needs to be thoroughly dried and sifted to eliminate as much of the seed from hay they ingest as possible. Not doing so will only add to the time you spend weeding.

There are other beneficial nutrients that can be added to the soil if necessary. They include lime, bone meal and nitrogen (to name a few). You need to remember, though, that all soil is not alike; meaning not all soil needs the same types of added nutrients.

I could take the time to explain how to test your soil and what to do with the results, but the following website does such a great job, I decided to pass it on to you for you to read.
http://www.gardeners.com/how-to/building-healthy-soil/5060.html

NOTE: While you may read/hear otherwise, having a special box or container to create compost is not necessary. There's nothing wrong with composting that way, but again…it's not an absolute must.

As for what you can compost, the following make great additions to the soil: **dead leaves, vegetables and vegetable peelings, coffee grounds, tea grounds, egg shells, oatmeal, bread, fruit and fruit peelings.**

Aeration means to allow air to circulate through something…in this case, your garden soil. Aeration is accomplished naturally by worms tunneling in, out and around the soil. Aeration is also accomplished when you hoe the dirt up between the rows of plants and around the plants (carefully so you don't uproot them). In other words, don't allow the soil in your garden to become packed down and harden. Aeration is also necessary to allow water to flow through the soil and get to the roots of your plants as efficiently and effectively as possible.

Water is the other essential for a healthy garden. We like to hope that nature will provide enough rainfall to keep the garden watered, but in most places this doesn't happen. That's why it is necessary to have a hydrant nearby—one close enough that a hose can easily reach every inch of garden space.

Companion planting is the act of planting vegetables together for optimal growth. Companion gardening cuts down on the amount of time you spend ridding your garden of pests and allows plants to give and take from each other. You can find a number of informative companion planting guides online—especially on PINTEREST.

Pest protection. Companion planting does help in the war against pests in your garden, but the self-sufficient farmer needs to understand that bugs, worms, slugs and other parasites do their best to make a meal out of your garden. Sevin® dust or spray is an old and reliable source of pest control. No, it's not all-natural, but it works. Other effective measures include spraying plants with a solution of dish soap and water, placing beer in shallow saucers of bowls in the garden, halved-grapefruits ward off slugs and scattering human hair on the garden wards off deer. If rabbits are real problem for you, you might need to place a chicken-wire fence around your garden to keep them out. Scarecrows do work, to some extent, as does tying aluminum pie tins to corn stalks (randomly) and putting black cats cut from wood or metal randomly along the edges of your garden.

Fresh garden lettuce

Cucumber plants ready for your garden

Chapter 4: Fruits, Herbs And Flowers

Now let's move on to fruits.

Strawberries are easy to grow and it doesn't take a lot of room to have enough for having fresh berries and for making jelly. Like asparagus, strawberries need to remain undisturbed other than to keep the weeds out and break off runners to thin the plants so they don't become over-crowded. Don't throw those runners away, though, you can give them away or sell them.

Blackberries, raspberries, blueberries and grapes all take more time and labor to establish, but if you have the room and your heart is set on it, then go for it.

Blueberries…they take a while to establish but are usually bountiful.

A couple of fruit trees can yield all the fruit your family will need for a year—including freezing or canning and making jelly—as long as the spring weather allows for blooming and germination and as long as you spray and care for your fruit trees in the proper way at the proper times.

It's fair to say that not all fruit trees thrive and produce in all planting zones. We all know the perils cold weather brings to the citrus trees and how one random late spring frost can destroy the year's peach crop in the Midwest. For this reason, it is important that you plant only those types of fruit trees that are hardy in your area.

Fruit trees do take up space, but I know several small farms that make the most of their land by growing fruit trees in the same pasture their sheep reside in. As long as the trees are limbed up (meaning no limbs hang low enough for the sheep to eat from them, both you and the sheep benefit. The trees provide a bit of shade, leaves to eat when they fall and even a bit of fruit that falls from the trees can be safely eaten. As for you, you get fresh fruit without having to waste time mowing around trees in an orchard.

If you don't mind mowing, fruit trees make a great welcoming committed on either side of your driveway or along the front of your yard. Planting them there saves room for other things elsewhere on the farm. Oh, and one last thing…if you make bees a part of your self-sufficiency, the bees are good for the fruit trees and the fruit trees are good for the bees' ability to make honey.

One definite plus for including fruit trees on your farm is the fact that all fruit can be eaten fresh as well as being preserved through canning and/or freezing for making jelly, pies and smoothies.

The final area we'll be discussing in regards to growing food for self-sufficiency is herbs and flowers.

Herbs and flowers can easily be grown around the outside of your house; bringing color, aroma and curb appeal to your home. The uses for herbs and flowers in regards to being self-sufficient are almost countless. You can use herbs for cooking, medicinal purposes, beauty treatments and to make your house smell clean and fresh.

Delicate Passion Vine

More specifically, fresh and dried herbs are a much tastier and healthier way to season food than salt is. Some of the most popular herbs to grow for using this way include: Dill, Basil, Sage, Rosemary, Mint, Thyme, Garlic, Cilantro, Parsley, Ginger, Fennell and Oregano.

Herbs used for medicinal purposes include those listed above as well as Lavender, Echinacea, Chamomile and Lemon Balm.
You can use most herbs fresh off the plant or you cut and dry them for later use.

It is important that you dry the herbs thoroughly before sealing into bags or jars. Not doing so will result in moldy leaves and nothing for you to use. One the leaves of some of your herbs have dried thoroughly, you can also seal 1-2 teaspoons into tea bags to make your own tea. Or if you prefer, you can brew loose tea in a tea ball or by straining it through a tea strainer.

The flowers from many herbs—including chamomile and mint—as well as more traditional flowers and plants like roses, violets and aloe, can be used in home-made soaps and added to mineral water or witch hazel for using to treat your skin and hair. If you aren't sure of how to 'create' these home-made treatments, there are a plethora of websites and magazine articles to get you started on your way to being at least somewhat self-sufficient in the beauty/cosmetics department.

Mint

Flowers and herbs can also be used to create crafts for yourself, your home, for selling and for gift giving. A few examples of this include:

Dried flower arrangements

Pressed flower artwork, bookmarks, cards

Herbal vinegars and oils

Herbal seasonings and teas

Home-made soaps and lotions

And just think…all that out of a little flower and herbal garden!

Conclusion

Let's recap what you've just read:

Being self-sufficient is labor-intensive, yet rewarding, good for your health and good for creating a true spirit of working together as a family.

Being self-sufficient can be achieved on a number of levels and the level at which you choose to be self-sufficient should fit within the amount of time, energy and effort you have to put towards it.

Being self-sufficient does *NOT* give you a license to cram everything imaginable onto a piece of ground and call yourself a farmer. Being self-sufficient means you make the most of your land in a way that is beneficial to you AND your land.

Being self-sufficient is a great way to save money, use your creative talents, gain a greater appreciation of what it takes to grow and produce the foods we eat.

Being self-sufficient requires a commitment and level-headedness on your part. Whether you look at it as a hobby, a job or a lifestyle…the level of commitment is there to be lived up to.

While this book emphasized how you can be self-sufficient by raising livestock, vegetables and fruits, the less-traditional agricultural ventures such as bees, nuts, fish ponds (aqua-farming) might also pique your interest. We've also not covered the more extreme levels of self-sufficiency such as solar energy, fiber arts and water collecting and filtering. Those, however, will have to be saved for another book. Besides…you've got enough to get you started, don't you agree?

Author Bio

 Darla Noble is a native of mid-Missouri where she lives with her husband of thirty-three years, John. Darla's love of writing began in the fourth grade; after meeting up and coming children's author, Judy Blume,
who, by the way, autographed Darla's copy of "Are you there, God...it's me, Margaret".
Darla's love for writing and family makes her work sought after in the Christian market, parenting and family resources and ghostwriting for educators and inspirational speakers.

THE MAGIC OF A
KITCHEN GARDEN
ORGANIC GARDENING FOR BEGINNERS

Health Learning Series
JD-Biz Publishing
Dueep J Singh and John Davidson

A BEGINNER'S GUIDE TO
HERB GARDENS
HOW TO GROW HERB GARDENS

The Garden Series
JD-Biz Publishing
Dueep J Singh and John Davidson

A BEGINNER'S GUIDE TO
HERB GARDENS

HERB GARDENING
IN YOUR HOME

The Healthy Garden Series
JD-Biz Publishing
Dueep J Singh and John Davidson

---

20 MOST BENEFICIAL
HERB PLANTS
FOR YOUR GARDEN
POPULAR HERBS FOR HEALTH AND CUISINE

HEALTHY GARDENING SERIES
JD-Biz Publishing
Dueep J Singh and John Davidson

A BEGINNER'S GUIDE TO
SUSTAINABLE GARDENING
PERMACULTURE TIPS FOR YOUR BACKYARD

HEALTHY GARDENING SERIES
JD-Biz Publishing
Dueep J Singh and John Davidson

A BEGINNER'S GUIDE TO
CACTI
HOW TO MAKE A CACTUS GARDEN

Health Learning Series
JD-Biz Publishing
Dueep J Singh and John Davidson

THE BEGINNER'S GUIDE TO
INDOOR & MINIATURE GARDENS
UNDERSTANDING INDOOR GARDENS, MOSS GARDENS, MINIATURE GARDENS AND GARDENS IN A BOTTLE

HEALTHY GARDENING SERIES
JD-Biz Publishing
Dueep J Singh and John Davidson

THE BEGINNER'S GUIDE TO
CITY GARDENING
SUSTAINABLE AND ORGANIC GARDENING IN LIMITED SPACE

HEALTHY GARDENING SERIES
JD-Biz Publishing
Dueep J Singh and John Davidson

THE BEGINNER'S GUIDE TO
RAISED BED GARDENING
GARDENING TIPS AND TECHNIQUES ON ORGANIC RAISED BED GARDENING

HEALTHY GARDENING SERIES
JD-Biz Publishing
Dueep J Singh and John Davidson

THE BEGINNER'S GUIDE TO
HOUSEPLANTS
EASY TIPS AND TECHNIQUES FOR GROWING HOUSEPLANTS IN YOUR HOME

HEALTHY GARDENING SERIES
JD-Biz Publishing
Dueep J Singh and John Davidson

A BEGINNER'S GUIDE TO
GROWING FRUIT TREES
GARDENING TIPS FOR GROWING FRUIT TREES FOR PLEASURE AND PROFIT

HEALTHY GARDENING SERIES
JD-Biz Publishing
Dueep J Singh and John Davidson

POT IT THERE
CONTAINER GARDENING
EVEN YOU CAN DO!

HEALTHY GARDENING SERIES
JD-Biz Publishing
Darla Noble and John Davidson

Health Learning Series

GRANDMA'S
NATURAL REMEDIES AND ANCIENT HERBAL BEAUTY RECIPES
Volume 1

HEALTH LEARNING SERIES
DULEEP J SINGH AND J DAVIDSON

GRANDMA'S
NATURAL REMEDIES AND ANCIENT HERBAL BEAUTY RECIPES

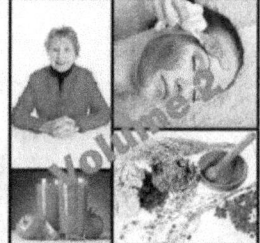

HEALTH LEARNING SERIES
DULEEP J SINGH AND J DAVIDSON

GRANDMA'S
NATURAL REMEDIES AND ANCIENT RECIPES
GRANDMA'S CURE FOR OBESITY
GRANDMA'S CURE FOR THE COMMON COLD
Volume 3

HEALTH LEARNING SERIES
DULEEP J SINGH AND J DAVIDSON

GRANDMA'S
NATURAL REMEDIES AND ANCIENT HERBAL RECIPES
Volume 4

HEALTH LEARNING SERIES
DULEEP J SINGH AND J DAVIDSON

GRANDMA'S
HERBAL LORE
ANCIENT HERBAL RECIPES AND REMEDIES
Volume 5

HEALTH LEARNING SERIES
DULEEP J SINGH AND J DAVIDSON

GRANDMA'S
ANCIENT BEAUTY REMEDIES
FROM HER KITCHEN
Volume 6

HEALTH LEARNING SERIES
DULEEP J SINGH AND J DAVIDSON

GRANDMA'S
EASY TO USE TIPS
IN THE KITCHEN AND OUTDOORS
Volume 7

HEALTH LEARNING SERIES
DULEEP J SINGH AND J DAVIDSON

GRANDMA'S
HOUSEHOLD HINTS AND RECIPES
USING TIME TESTED
ECONOMICAL TIPS IN YOUR HOME

75 Tips

Remove Stains

Peel Tomatoes   Perfect Pies

HEALTH LEARNING SERIES
DULEEP J SINGH AND J DAVIDSON

GRANDMA'S
NATURAL REMEDIES AND ANCIENT RECIPES

ALL 5 BOOKS IN 1

HEALTH LEARNING SERIES
DULEEP J SINGH AND J DAVIDSON

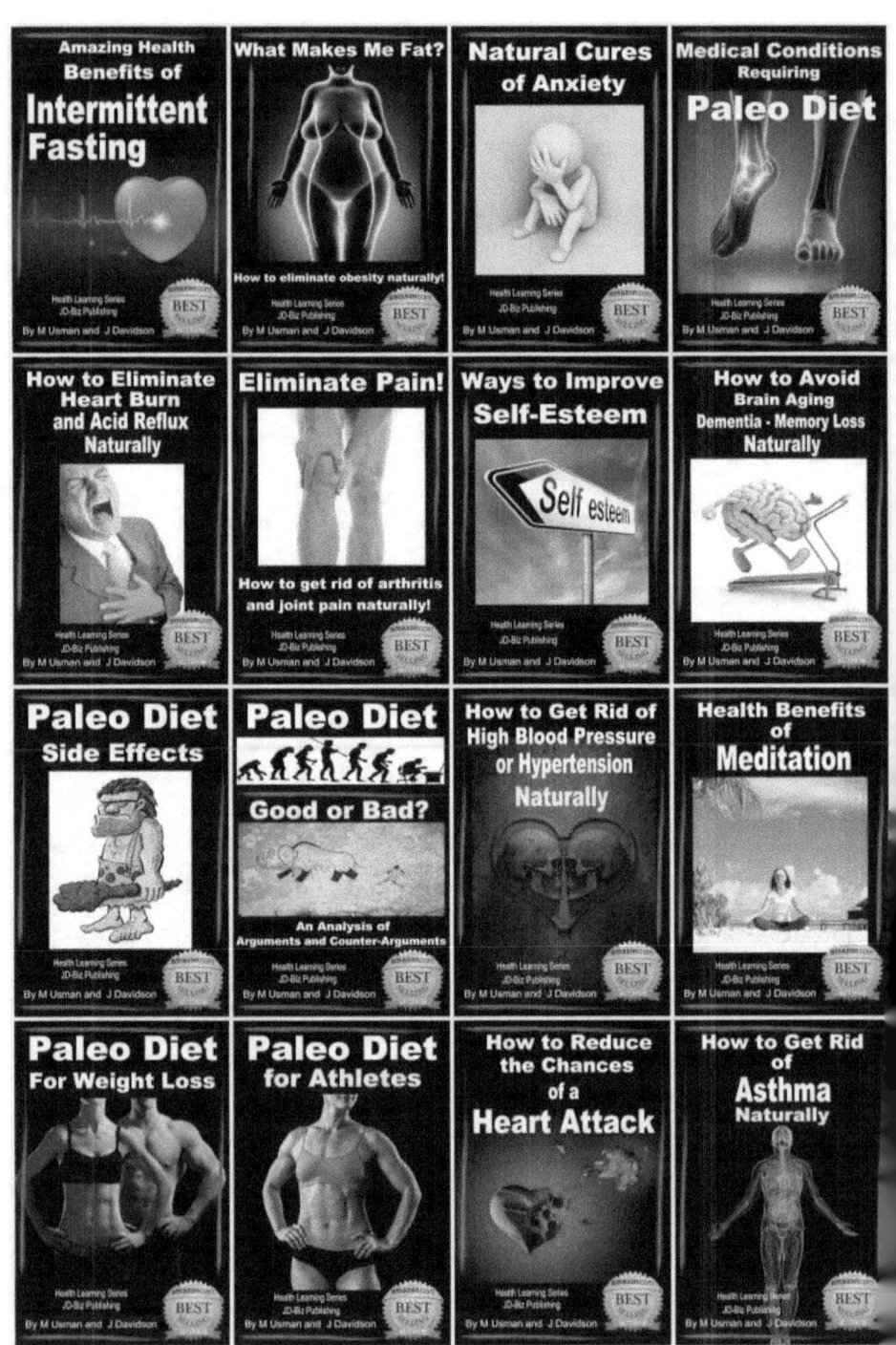

Amazing Animal Book Series

Chinchillas — Amazing Animal Books For Young Readers — By J Davidson plus J Rhynghburger — BEST

Beavers — For Kids — Amazing Animal Books For Young Readers — By J Davidson — BEST

Snakes — For Kids — Amazing Animal Books For Young Readers — By John Davidson and Nadeta Risso — BEST

Dolphins — For Kids — Amazing Animal Books For Young Readers — By John Davidson and Natalie Astor — BEST

Wolves — For Kids — Amazing Animal Books For Young Readers — By John Davidson and Virginia Forbes — BEST

Walruses — For Kids — Amazing Animal Books For Young Readers — By John Davidson and Kim Chase — BEST

Polar Bears — For Kids — Amazing Animal Books For Young Readers — By John Davidson and Kim Chase — BEST

Turtles — For Kids — Amazing Animal Books For Young Readers — By John Davidson and Natalie Astor — BEST

Bees — For Kids — Amazing Animal Books For Young Readers — By J Davidson and Jennifer Legacie — BEST

Frogs — For Kids — Amazing Animal Books For Young Readers — By John Davidson — BEST

Horses — For Kids — Amazing Animal Books For Young Readers — By John and Annalee Davidson — BEST

Monkeys — For Kids — Amazing Animal Books For Young Readers — By John and Annalee Davidson — BEST

Dinosaurs — For Kids — Amazing Animal Books For Young Readers — By John Davidson — BEST

Sharks — For Kids — Amazing Animal Books For Young Readers — By John Davidson — BEST

Whales — For Kids — Amazing Animal Books For Young Readers — By John Davidson — BEST

Spiders — For Kids — Amazing Animal Books For Young Readers — By John Davidson — BEST

Big Cats — For Kids — Amazing Animal Books For Young Readers — By John Davidson — BEST

Big Mammals of Yellowstone — For Kids — Amazing Animal Books For Young Readers — By John Davidson — BEST

Animals of Australia — Amazing Animal Books For Young Readers — By John Davidson and Shawn Vincent Wilson — BEST

Sasquatch - Yeti Abominable Snowman Bigfoot — For Kids — Amazing Animal Books For Young Readers — By John Davidson — BEST

Giant Panda Bears — For Kids — Amazing Animal Books For Young Readers — By John Davidson — BEST

Kittens — For Kids — Amazing Animal Books For Young Readers — By John Davidson — BEST

Komodo Dragons — For Kids — Amazing Animal Books For Young Readers — By Lisa Berry & John Davidson — BEST

Lady Bugs — For Kids — Amazing Animal Books For Young Readers — By Jean Hall & John Davidson — BEST

Animals of North America — For Kids — Amazing Animal Books For Young Readers — By John Davidson — BEST

Meerkats — For Kids — Amazing Animal Books For Young Readers — John Davidson and Lisa Berry

Birds of North America — For Kids — Amazing Animal Books For Young Readers — By John Davidson — BEST

Penguins — For Kids — Amazing Animal Books For Young Readers — Kim Chase & John Davidson — BEST

Hamsters — For Kids — Amazing Animal Books For Young Readers — John Davidson — BEST

Elephants — For Kids — Amazing Animal Books For Young Readers — Kim Chase & John Davidson

Learn To Draw Series

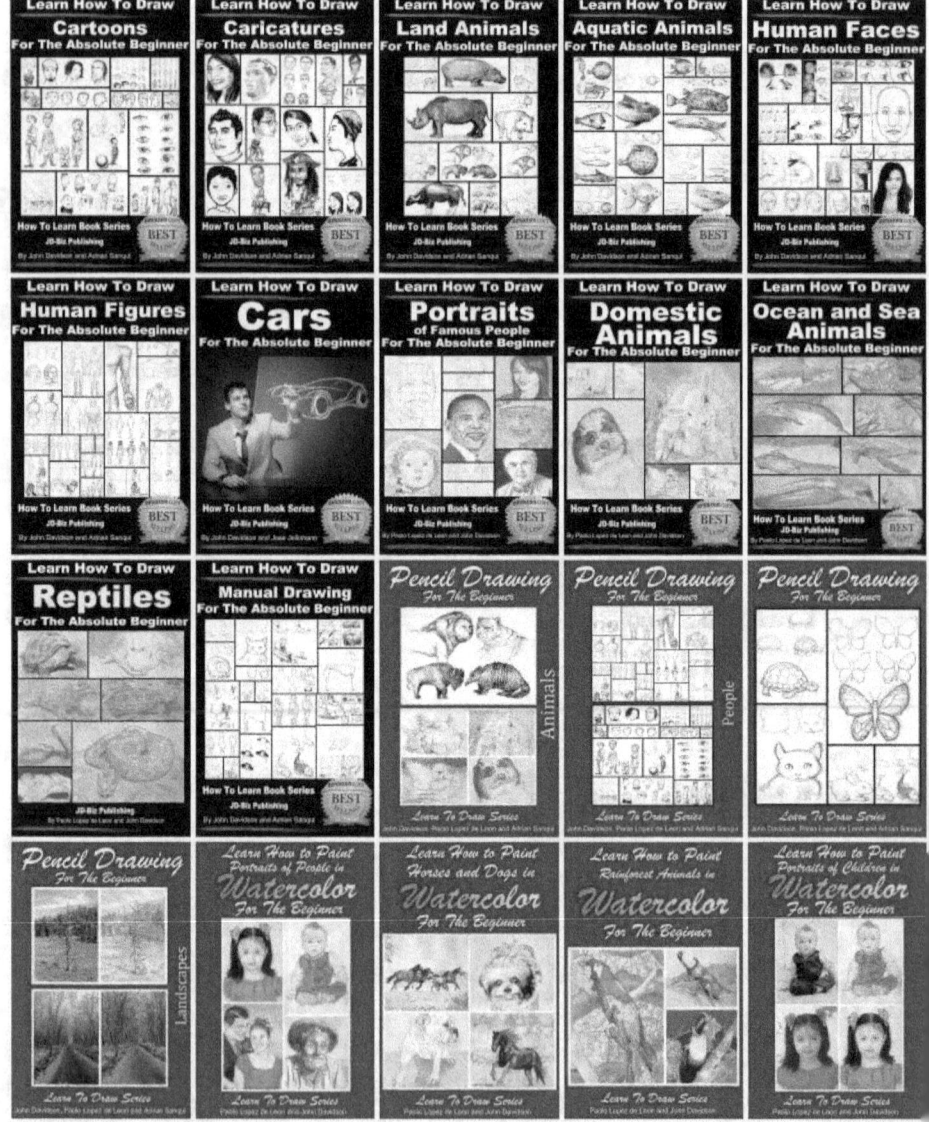

How to Build and Plan Books

Entrepreneur Book Series

Publisher

JD-Biz Corp

P O Box 374

Mendon, Utah 84325

http://www.jd-biz.com/

Mendon Cottage Books
P O Box 374, Mendon Utah 84325

www.ingramcontent.com/pod-product-compliance
Lightning Source LLC
Chambersburg PA
CBHW070719180526
45167CB00004B/1535